#TOXINS **tweet** Book01

140 Easy Tips to Reduce Your Family's Exposure to Environmental Toxins

By Dr. Laurel J. Standley

E-mail: info@thinkaha.com
20660 Stevens Creek Blvd., Suite 210
Cupertino, CA 95014

Copyright © 2011 by Dr. Laurel J. Standley

All rights reserved. No part of this book shall be reproduced, stored in a retrieval system, or transmitted by any means electronic, mechanical, photocopying, recording, or otherwise without written permission from the publisher.

Published by *THiNKaha*®, a Happy About® imprint
20660 Stevens Creek Blvd., Suite 210, Cupertino, CA 95014
http://thinkaha.com

First Printing: November 2011
Paperback ISBN: 978-1-61699-066-4 (1-61699-066-X)
eBook ISBN: 978-1-61699-067-1 (1-61699-067-8)
Place of Publication: Silicon Valley, California, USA
Paperback Library of Congress Number: 2011931460

Trademarks

All terms mentioned in this book that are known to be trademarks or service marks have been appropriately capitalized. Neither Happy About®, nor any of its imprints, can attest to the accuracy of this information. Use of a term in this book should not be regarded as affecting the validity of any trademark or service mark.

Warning and Disclaimer

Every effort has been made to make this book as complete and as accurate as possible. The information provided is on an "as is" basis. The author(s), publisher, and their agents assume no responsibility for errors or omissions. Nor do they assume liability or responsibility to any person or entity with respect to any loss or damages arising from the use of information contained herein.

Advance Praise

"Thomas Edison said 'The doctor of the future will...interest his patients in the...prevention of disease.' Well, the future is here! Dr. Laurel Standley offers us fresh, practical advice for taking control of protecting our health with short, easy to digest nuggets. Read them, share them, and live well!"
Dawn Suiter, Mom, Nutrition and Wellness Advocate

"I am so excited that Laurel has written this book. Laurel has a wonderful way of turning the most complicated, technical, scientific jargon into understandable terms. Add to that a wonderful sense of humor and a genuine love of her field and you have a winner. Laurel is my go-to scientist when I need a clearer understanding of what it means to 'be green.'"
Linda J. Brown, Linda Brown Coaching

"If the hype about toxins scares you, *#TOXINS tweet Book01* is here to put you on the right track. Laurel's comprehensive and easy-to-read presentation will show you what to look for and what to avoid. As a sustainable interior designer, *#TOXINS tweet Book01* is a resource I can reach for with confidence when designing healthy environments."
Donnalynn Polito, Leed Green Associate, Sustainable Interior Designer, http://www.ecolivingdesign.com

---

"The subject of environmental toxins can be complex and confusing for most of us. Laurel has a gift for explaining what they are, where they are, and the risks they pose to our health and our environment. Best of all, she offers practical advice for avoiding exposure."
Karen Janowski, Education Chair and Member of the Leadership Team, GreenTown Los Altos

Dedication

To my beloved family, Carolyn Standley and Lynn Taylor, for their support of my work on getting toxins out of people's lives. And my wise friends, Suzanne Eder, Judy Cohen, and Janet Teixeira, for their wisdom and guidance in helping me find my voice to write this book.

Dr. Laurel J. Standley

Acknowledgments

I wish to acknowledge all the incredible work being done by researchers, environmental activists, doctors, and moms' groups who envision a world without toxins in our food, our homes, and our bodies. I also want to thank the incredible people at Happy About for creating this opportunity and their wonderful suggestions for improving the text.

Dr. Laurel J. Standley

Why Did I Write This Book?

There is an overwhelming amount of information in the news and on the Internet these days about toxins in our environment. With lives packed full, it can be difficult to find the time to sift through this mass of information to figure out what is most important to you and your family's health. My goal in writing this book is to provide you with a simple guide packed with credible information and advice that you can readily apply in your daily life to reduce your exposure to the many toxic substances present in your environment and in the products you buy. It is my hope that as more of us choose safer products, manufacturers will respond by producing consumer goods using safer ingredients and cleaner manufacturing processes.

Dr. Laurel J. Standley *@Laurel_Standley*
http://clear-current.com
info@clear-current.com

140 Easy Tips to Reduce Your Family's Exposure to Environmental Toxins

Contents

Section I
Toxins in Our Home and
Outdoor Environments — 13

Section II
Health Issues Linked to
Environmental Exposures — 23

Section III
Hope for the Future — 43

Section IV
Food and Water — 49

Section V
Personal Care Products — 63

Section VI
Pregnancies and Small Children — 71

Section VII Household and Garage	79
Section VIII Indoor Pollution	93
Section IX Outdoor Environments and Garden	103
Section X Pets	111
Appendix A Abbreviations	115
Appendix B Resources and Guides	117
Appendix C Endnotes	121
About the Author	129

#TOXINS **tweet**

Section I: Toxins in Our Home and Outdoor Environments

Section 1

Toxins in Our Home and Outdoor Environments

As a society, we have produced tens of thousands of chemicals over the last several decades, and that number doesn't include chemicals created as contaminants during combustion or other processes.[1] Not all of these are toxic, but we currently know very little about how the majority of these chemicals affect our health. In this section, I provide an overview of the toxic chemicals we are exposed to in our daily lives, and the challenges we face in reducing our exposure to them.

Section I: Toxins in Our Home and Outdoor Environments

1

How many synthetic chemicals are there? More than 80,000 chemicals have been produced in the US since World War II.[2]

2

More than 2,000 synthetic chemicals are produced in volumes exceeding a million pounds each year in the US.[3]

> **3**
> Most of these chemicals have NOT been tested to determine their toxicity, particularly for neurological or reproductive problems.[4]

> **4**
> Body burdens: hundreds of chemicals have been detected in adults, children, and umbilical cords, indicating toxic exposure before birth.[5,6]

Section I: Toxins in Our Home and Outdoor Environments

5

Like the tip of an iceberg, that number doesn't include what we haven't measured yet or don't have the tools to measure.

6

Toxins we're exposed to include environmental pollutants, such as older pesticides, dioxins, PCBs, PAHs, and heavy metals.

7

We are also exposed to chemicals currently in use, such as newer pesticides, VOCs, formaldehyde, and gasoline.

8

There are many toxic chemicals present as pollutants in the air we breathe, water we drink, cook with, or swim in.

Section I: Toxins in Our Home and Outdoor Environments

9

We are exposed to toxic chemicals in some consumer goods like furniture, building materials, personal care, and household products.

10

How long does it take to clear a toxin from your body? Depends on whether it is fat soluble and if you are able to metabolize it.

11

Fat soluble chemicals such as dioxins, PCBs, older pesticides like DDT, and PAHs may take years to decades to clear from your body.[7,8]

Section I: Toxins in Our Home and Outdoor Environments

12

Take steps to avoid exposure to the fat soluble toxins as much as possible to keep your body burden lower.

13

Newer pesticides (organophosphates), plasticizers, and BPA are more rapidly cleared from your body.[9, 10]

14

Reducing your exposure to rapidly cleared chemicals can result in lower levels in your body within days.

15

Because of the widespread presence of toxins in the environment and consumer goods, it is not possible to avoid all exposures.

Section II: Health Issues Linked to Environmental Exposures

Section II

Health Issues Linked to Environmental Exposures

There are many health issues associated with our exposure to toxins because a wide variety of toxins act through different modes of toxicity in our bodies. It is difficult to prove beyond a shadow of a doubt that a specific toxic exposure is associated with human illness because it is unethical to directly expose people to toxins to see what will happen. We must therefore rely on other methods to determine whether toxic exposures are a problem for people, including studies on people exposed accidentally, animal testing, and/or epidemiological studies relating the differences between health outcomes for people with higher toxic exposure to those with lower or no toxic exposure. I have focused the key points in this section on health concerns with the strongest evidentiary links to toxic exposures.

Section II: Health Issues Linked to Environmental Exposures

16

Toxic environmental exposures have been linked to health issues like heart disease, cancer, asthma, and neurological impairment.[11]

17

Early life exposures to toxins are most damaging, with developing fetuses being very sensitive to these exposures.

18

Exposure to air pollution is related to many health issues: increased risk of strokes, asthma, miscarriage, and low birth weights.[12]

Section II: Health Issues Linked to Environmental Exposures

19

In addition to air pollution, asthma is linked to household items such as plastics (vinyl), formaldehyde (particle board), and ammonia.[13]

20

Exposure to pesticides, including herbicides, insecticides, and fungicides, is linked to a wide range of health effects in humans.[14]

21

Pesticides and reproduction = reduced fertility, abnormal sperm, asthma & Parkinson's.

22

Pesticides and exposure during pregnancy = low birth weight & birth defects, impaired learning, memory, attention span, and IQ.[15]

Section II: Health Issues Linked to Environmental Exposures

23

Pesticides and cancer = childhood leukemia; as well as brain, bladder, pancreatic, stomach, uterine, testicular, and prostate cancer.[16]

24

Exposure to heavy metals (e.g. lead, cadmium, mercury) is linked to many health issues, such as mental and cardiovascular diseases.[17]

25

Heavy metals and mental health = cognitive impairment, mental retardation, developmental delay, and ADD/ADHD.[18]

Section II: Health Issues Linked to Environmental Exposures

26

Heavy metals and reproduction = increased risk of miscarriage, reduced fertility, low birth weight, and altered sex ratios.[19]

27

Smoking is a well-known risk to human health; secondhand smoke is also considered a serious environmental hazard to non-smokers.

28

Secondhand smoking is linked to many cancers (e.g. breast, lung, and childhood brain), as well as asthma and coronary artery disease.

Section II: Health Issues Linked to Environmental Exposures

29

Secondhand smoking and reproduction/infant health = reduced fertility, miscarriage, low birth weight, and sudden infant death syndrome.[20]

30

Dioxins and PCBs, which accumulate in animal fats, are linked to a wide range of health issues for adults and their children.[21]

Section II: Health Issues Linked to Environmental Exposures

31
Dioxins/PCBs and reproduction/children's health = abnormal sperm, low birth weight, cognitive impairment, ADD/ADHD.[22]

32
Dioxins/PCBs and general health: immune suppression, coronary artery disease, Type II diabetes, hormonal changes, and various cancers.[23]

33

PAHs, created when food, wood, and other matter is burned, are linked to many cancers including lung, breast, bladder, and laryngeal.[24]

34

Formaldehyde from pressed wood, permanent press fabrics, and glues, is linked to leukemia, asthma, miscarriage, and reduced fertility.[25]

Section II: Health Issues Linked to Environmental Exposures

35

Endocrine disrupting chemicals (EDCs) include toxins such as DDT/DDE, PCBs, some plasticizers, BPA, and brominated flame retardants.[26]

36

EDCs mimic or interfere with our bodies' reproductive, neurological, and other hormone systems.[27]

37

The plasticizer DEHP, used to soften vinyl in toys, shower curtains, and faux leathers, has been linked to birth defects.[28]

Section II: Health Issues Linked to Environmental Exposures

38

BPA, a component of polycarbonate plastics, has been linked to miscarriages and altered neurodevelopment.[29]

39

Brominated flame retardants have been linked to miscarriages and altered neurodevelopment.[30]

40

Drinking water treatments to reduce water-borne disease may form disinfection byproducts = increased bladder and colorectal cancer risk.[31]

Section II: Health Issues Linked to Environmental Exposures

41

Solvents (e.g. gasoline, cleaners) have been linked to many health concerns, including bladder, breast, and childhood brain cancers.

42

Solvents and reproductive health = reduced fertility, miscarriage, and low birth weight.[32]

43

Solvents and other health issues = hepatitis, cirrhosis, contact dermatitis, ADD/ADHD, hyperactivity, and cognitive impairment.[33]

Section III: Hope for the Future

Section III

Hope for the Future

I realize the information in the previous two sections can be upsetting and too often people feel a bit overwhelmed by the many sources of toxins and the serious health effects their presence can have in our bodies. However, there are a few new studies showing that changing our personal environment can reduce our exposure and health outcomes from some of these toxic exposures.[34, 35, 36, 37] In this section, I discuss these studies and the hope they bring that we will be able to reduce the toxins present in our bodies and our environment.

Section III: Hope for the Future

44

Things can get better; intervention studies are showing health impacts and toxic body burdens can go down if exposures are reduced.

#TOXINS **tweet**

45

A Harvard study found that mortality associated with pollution was lowered in cities that improved their air quality over ten years.[38]

46

Fewer children were hospitalized for asthma when Atlanta cleared their air for the 1996 Olympics.[39]

Section III: Hope for the Future

47

Children switched to an organic diet cleared most organophosphate pesticide residues from their bodies within a day or two.[40]

48

People who switched from processed foods to non-packaged whole foods reduced BPA and plasticizers in their urine by more than half.[41]

Section IV: Food and Water

Section IV

Food and Water

Food and water can carry toxins that are either used intentionally for pest and disease control or are accumulated inadvertently from the environment and during cooking. Since many of these products are life essentials, we cannot avoid all toxins in these sources. However, there are choices you can make, such as selecting organic foods or cooking with stainless steel, to reduce your toxic exposure. In this chapter, I explain where many of the toxins associated with food and water come from, and the choices you can make to select less contaminated products.

Section IV: Food and Water

49
Choose organic produce and grains when possible (see produce guide to see the most and least contaminated produce).

50
In general, eating lower on the food chain means less exposure to toxins such as dioxins, PCBs, and organochlorine pesticides.

51

Because of biomagnification, fat soluble toxins are thousands to a million times more concentrated in animals than plants.

Section IV: Food and Water

52

Animals such as cattle, pigs, and chickens accumulate toxins in their fat—reduce consumption of such meats and dairy products.

53

Don't forget fat in cheese and butter; these products have been found to be contaminated by fat soluble toxins like PCBs and dioxins.[42]

54
Fish also accumulate toxins in their fats—choose safer fish and remove fat and skin before cooking (see safer fish guides).

55
However, fish contain omega-3 fats that are important for brain and heart health so just make sure you choose the safest ones to eat.

Section IV: Food and Water

56

Tuna, a lunchtime favorite, and mercury: chunk light contains the least, white albacore has more, and ahi tuna has the most.[43]

57

What about fish pills and flax oil instead of fish? Distilled fish oils are cleaner; flax oils are less beneficial than fats in fish.[44]

58

Even animals raised organically may absorb pollutants deposited from the air to their fields; eat less fat from these sources as well.[45]

59

When grilling meats, reduce carcinogen formation by soaking in antioxidant-rich marinades, avoiding sweet sauces, and minimizing char.[46]

Section IV: Food and Water

60

Burning any food, just like burning tobacco, creates carcinogens called PAHs; to reduce exposure, lightly toast and cut away char.

61

Cook in stainless steel, glass, or iron pans instead of plastic or non-stick to reduce exposure to phthalates and fluorinated chemicals.

62

Use glass or ceramic dishes instead of plastic to store food, especially if hot, moist or oily, which increases uptake of plasticizers.

63

Buy fresh, unpackaged foods rather than canned or packaged to reduce exposure to BPA and plasticizers.[47]

Section IV: Food and Water

64

In particular, avoid plastics with the recycling numbers 3, 6, and 7 on the bottom.

65

Contaminants can be present in drinking water sources, created during water treatment by utilities, or present in pipes within the home.

#TOXINS **tweet**

66

One source of water contaminants is the production of carcinogenic disinfection byproducts when water is treated to remove pathogens.

67

Lead may be present in older water pipes or solder connecting new pipes in homes; test water and replace pipes if contaminated.

Section IV: Food and Water

68

Bottled water is not necessarily safer than tap; water utilities face stricter guidelines and plastic water bottles may release toxins.[48, 49]

69

For safest drinking water, use filtered tap rather than bottled water; find the best filter using water treatment guides.

70

It is important to use filtered water when cooking as well to reduce exposures to toxins such as lead or disinfection byproducts.

71

Bottom line: avoid animal fats, eat more organic produce, cook or store food in glass, cast iron, or stainless steel.

Section V: Personal Care Products

Section V

Personal Care Products

Personal care products are not well regulated and may contain toxic ingredients or contaminants. Although not all ingredients of concern are listed on product labels, there are resources available to help you select safer products. In this section, I highlight several strategies you can use to reduce you and your family's exposure to toxic ingredients in hair and skin care products that you use each and every day.

Section V: Personal Care Products

72

Products used in personal care may present the greatest challenge in switching to safer products because of strong personal preferences.

73

Be aware that many personal care products contain ingredients of concern like EDCs and cancer-causing agents.

#TOXINS **tweet**

74
Using safer personal care products is important because we use these products directly on our bodies, increasing our risk of exposure.

75
To check the safety of your personal care products, scan labels for ingredients of concern and safety ratings.

Section V: Personal Care Products

76

Finding safer alternatives: check ratings for your products on the Skin Deep Database or Good Guide's product website.

77

Bring along EWG's wallet-sized guide of ingredients of concern when shopping for personal care products.

78

To reduce the cost of trying lots of products to find your favorites, trade with friends, and host a safer product testing party.

Section V: Personal Care Products

79

When you really love a product that contains ingredients of concern but have trouble finding a replacement, use it sparingly.

80

Skip using anti-microbial soaps, which work no better than regular soaps; wash hands and surfaces frequently or use sanitizers instead.[50]

81

Some sunscreen ingredients have health issues but it's still important to protect your skin from UV radiation; select safer products.

Section VI: Pregnancies and Small Children

Section VI

Pregnancies and Small Children

In the last decade, we have learned that fetuses are exposed to many if not all of the toxic exposures that are present in their mother's body. We are also learning that the effects of toxic exposures can be passed down to future generations through a process called epigenetics.[51] Small children, who are still developing their brains and reproductive systems, are also extremely vulnerable to toxic exposures. Protecting women in their child-bearing years and young children from these exposures is therefore essential for reducing birth defects and future health problems in our offspring. In this section, I offer suggestions for limiting toxic exposures to protect the health of our young.

Section VI: Pregnancies and Small Children

82
In-utero toxic exposures are of concern for health of developing babies; planning ahead for pregnancy may help reduce their risk.

83
While fat soluble toxins take years to clear, reducing exposure to phthalates, BPA, and newer pesticides can reduce Mom's toxic burden.

84

In particular, avoid use of and exposure to pesticides while pregnant or breast feeding.

85

Breast milk may contain many toxins from Mom's exposures but it is still important for babies to receive this for their overall health.[52]

Section VI: Pregnancies and Small Children

86

Use BPA-free labeled baby bottles and sippy cups; many U.S. manufacturers of baby bottles stopped using BPA in 2009.

87

Best option: use glass and stainless steel bottles instead of plastic, since other plastics may also leach hormonally-active toxins.

88

By mouthing their hands often, babies and small children are highly exposed to household toxins in dust and dirt; keep surfaces clean.

Section VI: Pregnancies and Small Children

89
Avoid soft plastic toys made from PVC (vinyl); they may contain hormonally-active plasticizers and the heavy metal lead.

90
Toys and children's jewelry may contain toxic metals like lead and cadmium; check consumer advisories and avoid cheap products.

91

Check websites for sources of safer toys and art supplies, as well as consumer advisory sites for warnings on existing products.

Section VII: Household and Garage

Section VII

Household and Garage

One of the most important things you can do to reduce your exposure to toxic substances is to clear products containing toxic materials from your home. Products used in the home and garage that may contain toxins include: cleaning products, furniture, flooring, electronics, and building materials. Although we are just beginning to learn about all of the potential sources of household toxins, I will provide suggestions for how you can reduce the use of toxic products in your home based on what we have learned thus far.

Section VII: Household and Garage

92

Choose simpler cleaning products like baking soda, vinegar, and castile soap to avoid toxins in some commercial products.

93

Minimize use of strong cleaners like bleach and ammonia, which can exacerbate health issues like asthma and eczema.

94

Household dust may contain many toxins, including EDCs; clean surfaces regularly with a damp cloth to avoid re-suspension of dust.[53]

Section VII: Household and Garage

95

Use a vacuum cleaner with a HEPA filter to remove smaller, inhalable particles from surfaces and air.

96

Synthetic fabrics in furniture and other items may contain toxic flame retardants.[54] Buy natural materials like cotton or wool.

#TOXINS **tweet**

97

New furniture made from particle board may release VOCs and formaldehyde;[55] replace with solid wood or older furniture.

Section VII: Household and Garage

98
Avoid use of vinyl (PVC) plastic items in your home, e.g. toys, shower curtains, faux leather, and window blinds.

99
Electronics may contain flame retardants, heavy metals, and PVC (vinyl) wire coverings; check guides for safer options.

100

Printers can emit fine particles;[56] place printer in another room if possible or away from your immediate work space.

101

Washing hands is not just good for reducing germs; it is also found to reduce exposure to toxic flame retardants for office workers.[57]

Section VII: Household and Garage

> **102**
> Garages and basements often contain a range of toxic products (e.g. gasoline, paints, pesticides) that outgas into your home's indoor air.

> **103**
> Clear out dangerous products you aren't using and dispose of them during your town's hazardous waste drive.

104

Working with toxic materials (pesticides, paints, or solvents)? Use protective gear, change clothes immediately, and wash them separately.

Section VII: Household and Garage

105

Ventilate well when working with solvents or gasoline in the garage or basement. Safest option: wear an organic vapor respirator.

106

Similarly, step away from the pump while you are filling your tank at the gas station.

107

Older homes may contain lead-based paints; major sources contributing to exposure are old wooden windows and peeling paint.[58]

Section VII: Household and Garage

108

Check guides before shopping to find safer alternatives for home products and building materials.

109

Bottom line: keeping all toxin-containing products out of your home is difficult; ventilate and clean frequently to reduce exposure.

Section VIII: Indoor Pollution

Section VIII

Indoor Pollution

Many people spend the majority of their time inside their home or in office environments. Research has shown that unless you live in a heavily urbanized area, indoor air may be more contaminated with pollutants than outdoor air.[59] In this section, I focus on important things that you can to do to improve indoor air quality in your home.

Section VIII: Indoor Pollution

110

Indoor air is often more contaminated than outdoor air; keeping toxic products out of your home is critical to protecting air quality.[60]

111

Newer, tightly sealed construction can stagnate indoor air: open windows occasionally to clear toxins and improve air quality.

112

To reduce pollution brought in from the outdoors, leave shoes at the door and have a carpet at the entrance to filter incoming dirt.[61]

Section VIII: Indoor Pollution

113

Radon exposure is the second leading cause of lung cancer; check EPA's radon map and test if you're located in a high concentration zone.[62]

114

Ventilate indoor sources of combustion (e.g. gas stoves, furnaces, water heaters), which may produce carbon monoxide and carcinogens.

115

Fireplaces and woodstoves are also sources of indoor air pollution like carbon monoxide and soot; make sure the smoke is exhausted properly.

116

Lighting candles and incense produce smoke that contains carcinogens;[63] soy and beeswax candles may be better than paraffin.

Section VIII: Indoor Pollution

117

Reduce emissions of toxins indoors from building materials like cabinets, carpets, and paint. Select safer products (see guides).

118

An attached garage can contaminate indoor air with toxic fumes; let exhaust clear before closing garage door and seal entrance to home.

#TOXINS **tweet**

119

Use HEPA filters on furnaces and fans to reduce indoor levels of small, inhalable particles that can carry toxins.

Section VIII: Indoor Pollution

120

In addition to reducing emissions of toxic fumes indoors, improve air quality indoors with house plants (3 per small room).[64]

121

The best indoor plants for clearing air are Boston ferns, Gerbera daisies, date and bamboo palms, rubber plants, and English ivies.[65]

122

Light "pollution" at night is linked to higher risk of breast cancer; use dark curtains in bedrooms, especially in well-lit urban areas.[66]

Section IX: Outdoor Environments and Garden

Section IX

Outdoor Environments and Garden

Enjoying the outdoors, whether in your backyard, or away from home, can be very beneficial to your health. Venturing out to exercise or a daily commute to work is an essential part of our lives, however, there are many sources of toxins present outside the home, including those used for pest control and the many chemicals present in air pollution. In this section, I provide you with suggestions for reducing your use of chemicals in your garden and also ways to reduce your exposure to pollutants present in the wider environment.

Section IX: Outdoor Environments and Garden

123

Location, location, location: air quality where you live can have a major impact on your health; check air quality in your area.

124

Air pollution is greatest near roadways during commute times and later in the day;[67] exercise early and commute before peak traffic.

125

Keep your car's air on recycle mode when driving in traffic, especially when following diesel trucks which emit the most toxins.[68]

126

Choose routes less traveled by vehicles to reduce your exposure to air pollution while walking, running, or biking.

Section IX: Outdoor Environments and Garden

127

Biking in high traffic areas? Channel your inner Darth Vader by wearing a respirator that filters fine particles (HEPA) and organic vapors.

128

Swimming in chlorinated pools results in exposure to carcinogenic disinfection byproducts.[69]

129

Reduce exposure to disinfection byproducts while swimming; seek a pool using less toxic disinfection and avoid drinking pool water.

130

There are many safer alternatives to pesticides. Check websites dedicated to organic or integrated pest management.

Section IX: Outdoor Environments and Garden

131

The best pest control approach is prevention; avoid stagnant water and leaks, clean up food spills quickly, and seal food in containers.

132

Urban soils can contain heavy metals like lead and other toxins; cover with clean soils in areas where children play.

133

When growing food near old (pre-1980s) homes and in cities, use clean soils added to raised beds; test the soil for lead.

134

Minimize use of biosolids (treated sewage) for growing food as they may contain toxins not removed during treatment.

Section X: Pets

Section X

Pets

Our furry companions are just as susceptible to toxins in the home and environment as we are. New studies are finding many of the same toxins in cats and dogs as those that have been detected in humans.[70, 71] Evidence is also emerging that some of our pets' health problems are associated with those toxic exposures.[72] In general, clearing your home of toxins can reduce both you and your pet's exposure to contaminants. In this section, I list additional suggestions for limiting exposure to toxins specifically for pets.

Section X: Pets

135

Dogs and cats are exposed to many of the same toxins as their humans, at times carrying higher body burdens than humans.[73]

136

Some dog breeds are more vulnerable to cancer, which has been linked to use of lawn chemicals such as weed-killers.[74]

137

Cat foods made with fish can be a major source of toxins; try to get them to eat other foods (and good luck if they prefer fish!).[75]

138

Cats, like infants, are also particularly exposed to household toxins that accumulate in dust and dirt on floors—vacuum often.

Section X: Pets

139

Some plastics leach reproductive toxins; use stainless steel or ceramic bowls to serve pets, especially if they're pregnant.

140

Birds are extremely sensitive to chemicals released when non-stick pans are heated to high temperatures; avoid use of these pans near birds.

Appendix A: Abbreviations

Note: "Chemical" is used in this book to mean synthetic, human-made substances rather than those present in nature.

ADD/ADHD	Attention Deficit (Hyperactivity) Disorder
BPA	Bisphenol A
DDT	For this book: includes the original pesticide, Dichlorodiphenyltrichloroethane, and its metabolites
DEHP	Diethyl Hexyl Phthalate, plasticizer
EDC	Endocrine Disrupting Chemical
EPA	Environmental Protection Agency
HEPA	High Efficiency Particulate Air filter
IQ	Intelligence Quotient
PAHs	Polycyclic Aromatic Hydrocarbons
PCBs	Polychlorinated Biphenyls
PVC	Polyvinyl Chloride
UV	Ultraviolet Light
VOC	Volatile Organic Compound

140 Easy Tips to Reduce Your Family's Exposure to Environmental Toxins

Appendix B: Resources and Guides

The following guides and resources provide additional information that can be useful in selecting safer products and reducing your exposure to toxins.

Food and Water
- EWG's dirty dozen & clean fifteen produce lists: http://www.ewg.org/foodnews/guide/
- Fish consumption guides:
  - http://www.nrdc.org/health/effects/mercury/guide.asp (mercury only)
  - http://apps.edf.org/documents/1980_pocket_seafood_selector.pdf (mercury, PCBs, and sustainability of resource)
  - Sushi guide: http://apps.edf.org/documents/8683_sushi_pocket.pdf
  - Fish oil pills: http://apps.edf.org/page.cfm?tagID=16536
- Water filtration products:
  - http://www.nsf.org/Certified/dwtu
  - http://bit.ly/NSF_Water_Treatment[1]

Personal Care Products
- Personal care products:
  - http://www.ewg.org/skindeep
  - http://www.goodguide.com
  - http://www.womensvoices.org/learn-more/
- Ingredients of concern wallet guide: http://www.ewg.org/files/EWG_cosmeticsguide.pdf

---

1. www.nsf.org/consumer/drinking_water/dw_treatment.asp?program=WaterTre

- Safer sunscreens: http://breakingnews.ewg.org/2011sunscreen/

Pregnancies and Small Children

- Safer toys:
  - http://www.healthystuff.org/departments/toys/product.using.php
  - http://www.cdc.gov/nceh/lead/recalls/toys.htm
- Children's food, clothes, and personal care products: http://www.goodguide.com

Household and Garage

- Better Basics for the Home: Simple Solutions for Less Toxic Living, Annie Berthold-Bond, Three Rivers Press, ©1999, 339 pp.
- Safer household cleaners: http://www.epa.gov/dfe
- Less-toxic electronics:
  - http://www.goodguide.com/categories/332304-cell-phones##products (cell phones only)
  - http://bit.ly/greenpeace-companies-line-up[2]
- Resources for selecting safer household products & building materials:
  - http://www.goodguide.com
  - http://www.greenguard.org
- Avoiding lead exposures: http://www.cdc.gov/nceh/lead/tips.htm
- Product recalls:
  - http://www.recalls.gov
  - http://www.cdc.gov/nceh/lead/Recalls/default.htm (lead, specifically)

---

2. www.greenpeace.org/international/en/campaigns/toxics/electronics/how-the-companies-line-up/

Indoor Pollution
- Radon info:
  - http://www.epa.gov/radon/pubs/citguide.html
  - http://www.epa.gov/radon/zonemap.html
    (locations of radon hotspots)

Outdoor Environments and Garden
- Air quality in your area: http://www.airnow.gov
- Safer pest control (organic or integrated pest management):
  http://www.ipm.ucdavis.edu/PMG/menu.homegarden.html

Pets
- Safer flea and tick control:
  http://www.greenpaws.org/_docs/GP_pocketguide.pdf
- Pet products:
  - http://www.healthystuff.org/departments/pets/product.using.php
  - http://www.goodguide.com/categories/308912-pet-food##products

Laboratories to contact if you want to have something tested for toxins:
- Contaminants in drinking water: search internet for '[Your State] Certified Laboratory Water Testing'
- Lead in paints, soils & other materials:
  http://www.epa.gov/lead/pubs/nllaplist.pdf
- Radon in indoor air: http://www.epa.gov/radon/pubs/citguide.html

140 Easy Tips to Reduce Your Family's Exposure to Environmental Toxins

Appendix C: Endnotes

Section I: Toxins in Our Home and Outdoor Environments

1. U.S. Environmental Protection Agency. "TSCA Chemical Substance Inventory." *Environmental Protection Agency.* Last updated on August 5, 2011. http://tinyurl.com/3uke5gj.[3]
2. Ibid.
3. "High Production Volume (HPV) Challenge." *Environmental Protection Agency.* Last updated on January 31, 2011. http://www.epa.gov/hpv.
4. Judson, Richard, Ann Richard, David Dix, Keith Houck, Matthew Martin, Robert Kavlock, Vicki Dellarco, Tala Henry, Todd Holderman, Philip Sayre, Shirlee Tan, Thomas Carpenter, and Edwin Smith. "The Toxicity Data Landscape for Environmental Chemicals." *Environmental Health Perspectives* 117, no.5 (2009): 685–695.
5. "The Environment: BodyBurden." Environmental Working Group. Accessed August 2011. http://www.ewg.org/sites/bodyburden1/es.php.
6. Department of Health and Human Services, Centers for Disease Control and Prevention. *Third National Report on Human Exposure to Environmental Chemicals.* Atlanta, GA: National Center for Environmental Health, Division of Laboratory Services: July 2005.
7. Ogura, Isamu. "Half-life of each dioxin and PCB congener in the human body." *Organohalogen Compounds* 66, (2004): 3329–3337.
8. Noren, Koldu, and Dalva Melronyte. "Certain Organochlorine and Organobromine Contaminants in Swedish Human Milk in Perspective of Past 20–30 Years. *Chemosphere* 40, no. 9–11 (May–June 2000): 1111–1123.
9. Lu, Chensheng, Kathryn Toepel, Rene Irish, Richard A. Fenske, Dana B. Barr, and Roberto Bravo. "Organic Diets Significantly Lower Children's Dietary Exposure to Organophosphorus Pesticides." *Environmental Health Perspectives* 114, (2006): 260–263.

---

3. www.epa.gov/oppt/existingchemicals/pubs/tscainventory/basic.html#background

10 Rudel, Ruthann, Janet Gray, Connie Engel, Teresa Rawsthorne, Robin Dodson, Janet Ackerman, Jeanne Rizzo, Janet Nudelman, and Julia Green Brody. "Food Packaging and Bisphenol A and Bis-(2-Ethylhexyl)Phthalate Exposure: Findings from a Dietary Intervention." *Environmental Health Perspectives* 119, (2011): 914–920.

Section II: Health Issues Linked to Environmental Exposures

11 "CHE Toxicant and Disease Database." *The Collaborative on Health and the Environment.* Accessed August 2011. http://www.healthandenvironment.org/tddb.

12 Ibid.
13 Ibid.
14 Ibid.
15 Ibid.
16 Ibid.
17 Ibid.
18 Ibid.
19 Ibid.
20 Ibid.
21 Ibid.
22 Ibid.
23 Ibid.
24 Ibid.
25 Ibid.
26 Ibid.
27 Ibid.
28 Ibid.
29 Ibid.

30 Ibid.

31 Ibid.

32 Ibid.

33 Ibid.

Section III: Hope for the Future

34 Laden, Francine, Joel Schwartz, Frank Speizer, and Douglas Dockery. "Reduction in Fine Particulate Air Pollution and Mortality: Extended Follow-up of the Harvard Six Cities Study." *American Journal of Respiratory and Critical Care Medicine* 173, (2006): 667–672.

35 Friedman, Michael, Kenneth Powell, Lori Hutwagner, LeRoy Graham, and W. Gerald Teague. "Impact of Changes in Transportation and Commuting Behaviors During the 1996 Summer Olympic Games in Atlanta on Air Quality and Childhood Asthma." *Journal of the American Medical Association* 285, (2001): 897–905.

36 Lu, Chensheng, Kathryn Toepel, Rene Irish, Richard A. Fenske, Dana B. Barr, and Roberto Bravo. "Organic Diets Significantly Lower Children's Dietary Exposure to Organophosphorus Pesticides." *Environmental Health Perspectives* 114, (2006): 260–263.

37 Rudel, Ruthann, Janet Gray, Connie Engel, Teresa Rawsthorne, Robin Dodson, Janet Ackerman, Jeanne Rizzo, Janet Nudelman, and Julia Green Brody. "Food Packaging and Bisphenol A and Bis-(2-Ethylhexyl)Phthalate Exposure: Findings from a Dietary Intervention." *Environmental Health Perspectives* 119, (2011): 914–920.

38 Laden, Francine, Joel Schwartz, Frank Speizer, and Douglas Dockery. "Reduction in Fine Particulate Air Pollution and Mortality: Extended Follow-up of the Harvard Six Cities Study." *American Journal of Respiratory and Critical Care Medicine* 173, (2006): 667–672.

39 Friedman, Michael, Kenneth Powell, Lori Hutwagner, LeRoy Graham, and W. Gerald Teague. "Impact of Changes in Transportation and Commuting Behaviors During the 1996 Summer Olympic Games in

Atlanta on Air Quality and Childhood Asthma." *Journal of the American Medical Association* 285, (2001): 897–905.

40 Lu, Chensheng, Kathryn Toepel, Rene Irish, Richard A. Fenske, Dana B. Barr, and Roberto Bravo. "Organic Diets Significantly Lower Children's Dietary Exposure to Organophosphorus Pesticides." *Environmental Health Perspectives* 114, (2006): 260–263.

41 Rudel, Ruthann, Janet Gray, Connie Engel, Teresa Rawsthorne, Robin Dodson, Janet Ackerman, Jeanne Rizzo, Janet Nudelman, and Julia Green Brody. "Food Packaging and Bisphenol A and Bis-(2-Ethylhexyl)Phthalate Exposure: Findings from a Dietary Intervention." *Environmental Health Perspectives* 119, (2011): 914–920.

Section IV: Food and Water

42 Kalantzi, Olga-Ioanna, Ruth Alcock, Paul Johnston, David Santillo, Ruth Stringer, Gareth Thomas, and Kevin Jones. "The Global Distribution of PCBs and Organochlorine Pesticides in Butter." *Environmental Science and Technology* 35, (2001): 1013–1018.

43 "Consumer Guide to Mercury in Fish." Natural Resources Defense Council. Accessed August 2011. **http://tinyurl.com/29wws3**.[4]

44 Wilkinson, Paul, Clare Leach, Eric Ah-Sing, Nahed Hussain, George Miller, D. Joe Millward, and Bruce Griffin. "Influence of α-Linolenic Acid and Fish-Oil on Markers of Cardiovascular Risk in Subjects With an Atherogenic Lipoprotein Phenotype." *Atherosclerosis* 181, (2005): 115–124.

45 Lorber, Matthew, David Cleverly, John Schaum, Linda Phillips, Greg Schweer, and Timothy Leighton. "Development and Validation of an Air-to-Beef Food Chain Model for Dioxin-Like-Compounds." *The Science of the Total Environment* 156, (1994): 39–65.

46 Smith, J. Scott, F Ameri, and Priyadarshini Gadgil. "Effects of Marinades on the Formation of Heterocyclic Amines in Grilled Beef Steaks." *Journal of Food Science* 73, (2008): T100–T105.

---

4. www.nrdc.org/health/effects/mercury/guide.asp

47 Rudel, Ruthann, Janet Gray, Connie Engel, Teresa Rawsthorne, Robin Dodson, Janet Ackerman, Jeanne Rizzo, Janet Nudelman, and Julia Green Brody. "Food Packaging and Bisphenol A and Bis-(2-Ethylhexyl)Phthalate Exposure: Findings from a Dietary Intervention." *Environmental Health Perspectives* 119, (2011): 914–920.

48 Gleick, Peter H. Bottled and Sold: The Story Behind Our Obsession with Bottled Water. Washington, DC: Island Press, 2010.

49 Leiba, Nneka, Sean Gray, and Jane Houlihan. "2011 Bottled Water Scorecard." *Environmental Working Group*, (2011), http://tinyurl.com/2bouvl8.[5]

Section V: Personal Care Products

50 Aiello, Allison, Rebecca Coulborn, Vanessa Perez, and Elaine Larson. "Effect of Hand Hygiene on Infectious Disease Risk in the Community Setting: A Meta-Analysis." *American Journal of Public Health* 98, (2008): 1372–1381.

Section VI: Pregnancies and Small Children

51 Perera, Frederica, Julie Herbstman. "Prenatal Environmental Exposures Epigenetics, and Disease." Reproductive Toxicology 31, No. 3 (2011): 363–373.

52 Solomon, Gina, and Pilar Weiss. "Chemical Contaminants in Breast Milk: Time Trends and Regional Variability." *Environmental Health Perspectives* 110, (2002): A339–A347.

Section VII: Household and Garage

53 Rudel, Ruthann, David Camann, John Spengler, Leo Korn, and Julia Brody. "Phthalates, Alkylphenols, Pesticides, Polybrominated Diphenyl Ethers, and Other Endocrine-Disrupting Compounds in Indoor Air and Dust." *Environmental Science and Technology* 37, (2003): 4543–4553.

---

5. static.ewg.org/reports/2010/bottledwater2010/pdf/2011-bottledwater-scorecard-report.pdf

54 Webster, Thomas, Stuart Harrad, James Millette, R. David Holbrook, Jeffrey Davis, Heather Stapleton, Joseph Allen, Michael McClean, Catalina Ibarra, Mohamed Abou-Elwafa Abdallah, and Adrian Covaci. "Identifying Transfer Mechanisms and Sources of Decabromodiphenyl Ether (BDE 209) in Indoor Environments Using Environmental Forensic Microscopy." *Environmental Science and Technology* 43, (2009): 3067–3072.

55 Brown, Steve. "Chamber Assessment of Formaldehyde and VOC Emissions from Wood-Based Panels." *Indoor Air* 9, (1999): 209–215.

56 McGarry, Peter, Lidia Morawska, Congrong He, Rohan Jayaratne, Matthew Falk, Quang Tran, and Hao Wang (2011), "Exposure to Particles from Laser Printers Operating within Office Workplaces." *Environmental Science and Technology* 45, (2011): 6444–6452.

57 Watkins, Deborah, Michael McClean, Alicia Fraser, Janice Weinberg, Heather Stapleton, Andreas Sjödin. "Exposure to PBDEs in the Office Environment: Evaluating the Relationship Between Dust, Handwipes, and Serum." Environmental Health Perspectives, (June 30, 2011).

58 Lead in Paint, Dust, and Soil." U.S. Environmental Protection Agency. Last modified August 22, 2011. <http://www.epa.gov/lead/>.

Section VIII: Indoor Pollution

59 "The Inside Story: A Guide to Indoor Air Quality." United States Environmental Protection Agency. Accessed August 2011. <http://www.epa.gov/iaq/pubs/insidestory.html#Intro1>.

60 Ibid.

61 Nishioka, Marcia, Hazel Burkholder, and Marielle Brinkman. "Distribution of 2,4-Dichlorophenoxyacetic Acid in Floor Dust throughout Homes Following Homeowner and Commercial Lawn Applications: Quantitative Effects of Children, Pets, and Shoes." *Environmental Science and Technology* 33, (1999): 1359–1365.

62 "A Citizen's Guide to Radon." United States Environmental Protection Agency. Accessed August 2011. http://www.epa.gov/radon/pubs/citguide.html, downloaded August 2011).

63 Orecchio, Santino. "Polycyclic Aromatic Hydrocarbons (PAHs) in Indoor Emission from Decorative Candles." *Atmospheric Environment* 45, (2011): 1888–1895.

64 Wolverton, B. C. *How to Grow Fresh Air: 50 Houseplants that Purify Your Home or Office.* New York: Penguin Books, 1997.

65 Ibid.

66 Blask, David, George Brainard, Robert Dauchy, John Hanifin, Leslie Davidson, Jean Krause, Leonard Sauer, Moises Rivera-Bermudez, Margarita Dubocovich, Samar Jasser, Darin Lynch, Mark Rollag, and Frederick Zalatan. "Melatonin-Depleted Blood from Premenopausal Women Exposed to Light at Night Stimulates Growth of Human Breast Cancer Xenografts in Nude Rats." *Cancer Research* 65, (2005): 11,174–11,184.

Section IX: Outdoor Environments and Garden

67 Janssen, Nicole, Patricia van Vliet, Francée Aarts, Hendrik Harssema, and Bert Brunekreef. "Assessment of Exposure to Traffic Related Air Pollution of Children Attending Schools Near Motorways." *Atmospheric Environment* 35, (2001): 3875–3884.

68 Cal/EPA's Office of Environmental Health Hazard Assessment and The American Lung Association of California. "Health Effects of Diesel Exhaust." Accessed August 2011. http://tinyurl.com/3h2p5ov.[6]

69 Cardador, Maria, and Mercedes Gallego. "Haloacetic Acids in Swimming Pools: Swimmer and Worker Exposure." *Environmental Science and Technology* 45, (2011): 5783–5790.

---

6. oehha.ca.gov/public_info/facts/pdf/diesel4-02.pdf

Section X: Pets

70. "Pets for the Environment." Environmental Working Group. Accessed August 2011. http://www.ewg.org/PetsfortheEnvironment.

71. Dye, Janice, Marta Venier, Lingyan Zhu, Cynthia Ward, Ronald Hites, and Linda Birnbaum. "Elevated PBDE Levels in Pet Cats: Sentinels for Humans?" *Environmental Science and Technology* 41, (2007): 6350–6356.

72. Kelsey, Jennifer, Antony Moore, and Lawrence Glickman. "Epidemiologic Studies of Risk Factors for Cancer in Pet Dogs." *Epidemiologic Reviews* 20, (1998): 204–217.

73. "Pets for the Environment." Environmental Working Group. Accessed August 2011. http://www.ewg.org/PetsfortheEnvironment.

74. Kelsey, Jennifer, Antony Moore, and Lawrence Glickman. "Epidemiologic Studies of Risk Factors for Cancer in Pet Dogs." *Epidemiologic Reviews* 20, (1998): 204–217.

75. Dye, Janice, Marta Venier, Lingyan Zhu, Cynthia Ward, Ronald Hites, and Linda Birnbaum. "Elevated PBDE Levels in Pet Cats: Sentinels for Humans?" *Environmental Science and Technology* 41, (2007): 6350–6356.

About the Author

Dr. Laurel J. Standley, Principal of Clear Current LLC (http://Clear-Current.com), is an environmental consultant with over twenty years experience in environmental chemistry and policy. She received her B.S. in Chemistry from California Polytechnic State University, a Ph.D. in Chemical Oceanography from Oregon State University, and a M.A. in Urban Affairs and Public Policy from the University of Delaware.

Laurel has a deep personal interest in educating the public about reducing personal exposure to environmental toxins. She works with clients to achieve sustainable water use and transition toward green alternatives in product use and manufacturing. Her past projects have included presenting public workshops on reducing exposure to toxins, investigating the use of alternative products to reduce human exposure to endocrine disrupting chemicals, developing an innovative tool for New York City to protect their drinking water resources from contamination, and building a business strategy for a large nonprofit to engage corporations in protecting water resources in biodiversity hotspots.

Other Books in the THiNKaha Series

The THiNKaha book series is for thinking adults who lack the time or desire to read long books, but want to improve themselves with knowledge of the most up-to-date subjects. THiNKaha is a leader in timely, cutting-edge books and mobile applications from relevant experts that provide valuable information in a fun, Twitter-brief format for a fast-paced world.

They are available online at http://thinkaha.com or at other online and physical bookstores.

1. *#BOOK TITLE tweet Book01:* 140 Bite-Sized Ideas for Compelling Article, Book, and Event Titles by Roger C. Parker
2. *#BUSINESS SAVVY PM tweet Book01:* Project Management Mindsets, Skills, and Tools for Generating Successful Business Results by Cinda Voegtli
3. *#COACHING tweet Book01:* 140 Bite-Sized Insights On Making A Difference Through Executive Coaching by Sterling Lanier
4. *#CONTENT MARKETING tweet Book01:* 140 Bite-Sized Ideas to Create and Market Compelling Content by Ambal Balakrishnan
5. *#CORPORATE CULTURE tweet Book01:* 140 Bite-Sized Ideas to Help You Create a High Performing, Values Aligned Workplace that Employees LOVE by S. Chris Edmonds
6. *#CORPORATE GOVERNANCE tweet Book01:* How Corporate Governance Adds Value to Your Business! by Brad Beckstead
7. *#CROWDSOURCING tweet Book01:* 140 Bite-Sized Ideas to Leverage the Wisdom of the Crowd by Kiruba Shankar and Mitchell Levy
8. *#CULTURAL TRANSFORMATION tweet Book01:* Cutting-Edge Advice from Today's Global Business Leaders by Melissa Lamson
9. *#DEATHtweet Book01:* A Well-Lived Life through 140 Perspectives on Death and Its Teachings by Timothy Tosta

10. *#DEATH tweet Book02:* 140 Perspectives on Being a Supportive Witness to the End of Life by Timothy Tosta
11. *#DIVERSITYtweet Book01:* Embracing the Growing Diversity in Our World by Deepika Bajaj
12. *#DREAMtweet Book01:* Inspirational Nuggets of Wisdom from a Rock and Roll Guru to Help You Live Your Dreams by Joe Heuer
13. *#ENTRYLEVELtweet Book01:* Taking Your Career from Classroom to Cubicle by Heather R. Huhman
14. *#ENTRY LEVEL tweet Book02:* Relevant Advice for Students and New Graduates in the Day of Social Media by Christine Ruff and Lori Ruff
15. *#EXPERT EXCEL PROJECTS tweet:* Taking Your Excel Project From Start To Finish Like An Expert by Larry Moseley
16. *#GOOGLE+ tweet Book01:* Getting to Know Google's New Social Network by Julio Ojeda-Zapata
17. *#IT OPERATIONS MANAGEMENT tweet Book01:* Managing Your IT Infrastructure in The Age of Complexity by Peter Spielvogel, Jon Haworth, Sonja Hickey
18. *#JOBSEARCHtweet Book01:* 140 Job Search Nuggets for Managing Your Career and Landing Your Dream Job by Barbara Safani
19. *#LEADERSHIPtweet Book01:* 140 Bite-Sized Ideas to Help You Become the Leader You Were Born to Be by Kevin Eikenberry
20. *#LEADS to SALES tweet Book01:* Creating Qualified Business Leads in the 21st Century by Jim McAvoy
21. *#LEAN SIX SIGMA tweet Book01:* Business Process Excellence for the Millennium by Dr. Shree R. Nanguneri
22. *#LEAN STARTUP tweet Book01:* 140 Insights for Building a Lean Startup! by Seymour Duncker
23. *#MILLENNIALtweet Book01:* 140 Bite-Sized Ideas for Managing the Millennials by Alexandra Levit

24. *#MOJOtweet:* 140 Bite-Sized Ideas on How to Get and Keep Your Mojo by Marshall Goldsmith
25. *#MY BRAND tweet Book01:* A Practical Approach to Building Your Personal Brand—140 Characters at a Time by Laura Lowell
26. *#OPEN TEXTBOOK tweet Book01:* Driving the Awareness and Adoption of Open Textbooks by Sharyn Fitzpatrick
27. *#PARTNER tweet Book01:* 140 Bite-Sized Ideas for Succeeding in Your Partnerships by Chaitra Vedullapalli
28. *#PLAN to WIN tweet Book01:* Strategic Territory and Account Planning by Ron Snyder and Eric Doner
29. *#PRESENTATION tweet Book01:* 140 Ways to Present with Impact by Wayne Turmel
30. *#PRIVACY tweet Book01:* Addressing Privacy Concerns in the Day of Social Media by Lori Ruff
31. *#PROJECT MANAGEMENT tweet Book01:* 140 Powerful Bite-Sized Insights on Managing Projects by Guy Ralfe and Himanshu Jhamb
32. *#QUALITYtweet Book01:* 140 Bite-Sized Ideas to Deliver Quality in Every Project by Tanmay Vora
33. *#RISK MANAGEMENT tweet Book01:* Proactive Risk Management: Taming Alligators by Cinda Voegtli & Laura Erkeneff
34. *#SCRAPPY GENERAL MANAGEMENT tweet Book01:* Practical Practices for Magnificent Management Results by Michael Horton
35. *#SOCIAL MEDIA PR tweet Book01:* 140 Bite-Sized Ideas for Social Media Engagement by Janet Fouts
36. *#SOCIALMEDIA NONPROFIT tweet Book01:* 140 Bite-Sized Ideas for Nonprofit Social Media Engagement by Janet Fouts with Beth Kanter
37. *#SPORTS tweet Book01:* What I Learned from Coaches About Sports and Life by Ronnie Lott with Keith Potter
38. *#STANDARDS tweet Book01:* 140 Bite-Sized Ideas for Winning the Industry Standards Game by Karen Bartleson

39. *#SUCCESSFUL CORPORATE TRAINING tweet Book01:* Profitable Training by Optimizing your Customer and Partner Education Organization by Mitchell Levy and Terry Lydon

40. *#TEAMWORK tweet Book01:* Lessons for Leading Organizational Teams to Success 140 Powerful Bite-Sized Insights on Lessons for Leading Teams to Success by Caroline G. Nicholl

41. *#THINKtweet Book01:* Bite-Sized Lessons for a Fast Paced World by Rajesh Setty

42. *#TOXINS tweet Book01:* 140 Easy Tips to Reduce Your Family's Exposure to Environmental Toxins by Laurel J. Standley Ph.D.

THiNK Continuity™ Training/Learning Program

THiNK Continuity™ delivers high-quality, cost-effective continuous learning in easy-to-understand, worthwhile, and digestible chunks. Fifteen minutes with a *THiNKaha®* book will allow the reader to have one or more "aha" moments. An hour and a half monthly with a THiNK Continuity program will allow the learner to have an opportunity to truly digest the topic being covered.

Offered online and/or in person, these engaging programs feature gurus (ours and yours) on such relevant topics as Leadership, Management, Sales, Marketing, Work-Life Balance, Project Management, Social Media and Networking, Presentation Skills, and other topics of your choosing. The "learning" audience, whether it is clients, employees or partners, can now experience high-quality learning that will enhance your brand value and empower your company as a thought leader. This program fits a real need where time and the high cost of developing custom content are no longer an option for every organization.

Just **THiNK**...

- **C**ontinuous Employee/Client/Prospect Learning
- **O**ngoing Thought Leadership Development
- **N**otable Experts Presenting on Relevant Topics
- **T**ime Your Attendees Can Afford—15 min. to 2 hrs/mth
- **I**nformation Delivered in Digestible Chunks
- **N**ame the Topic—We Help You Provide Expert Best Practices
- **U**nderstand and Implement the Takeaways
- **I**nternal Expertise Shared Externally
- **T**raining/Prospecting Cost Decreases, Effectiveness Increases
- **Y**ou Win, They Win!

CPSIA information can be obtained at www.ICGtesting.com
Printed in the USA
BVOW031218131111

275979BV00006B/1/P